THE 48 LAWS OF POWER IN PRACTICE

The 3 most powerful laws and the
4 indispensible power principles.

JON WATERLOW
ANDREA DOMENICHINI

Designed and typeset by

Jaime Kavanagh

*Using Garamond 11 point for the main body and quotes, as
well as Dante
for various headings and Didot
for the dedications. Bodoni
Ornaments were used
for embellishment
and seperation.*

© 2017, Jon Waterlow & Andrea Domenichini
Cover, Illustrations: Jaime Kavanagh

Publisher: tredition

ISBN

Paperback   978-3-7323-8712-0

Hardcover   978-3-7323-8713-7

eBook   978-3-7323-8714-4

INTRODUCTION

There's a reason why you opened this book. You know there is, but the chances are you're not entirely comfortable admitting to yourself what it is. It's not because *The 48 Laws of Power* has sold millions of copies or that it's the most-requested book in America's prisons. It's something much more personal than that.

The 48 Laws of Power touches more than a few sensitive spots in our psyches; it forces us to reevaluate our relationships, and to start asking difficult questions about how we've been told society works. Of course, for some people this is simply too disturbing and destabilising. But while they might prefer to remain peacefully asleep at the wheel of their lives, we can face those questions and make real, lasting changes.

Perhaps you want to become a modern Machiavelli. Perhaps you want to escape the daily grind and realise your true potential and your dreams. Or maybe you're just tired of finding yourself the victim of other people's games, tossed about by power-plays that you don't realise are happening until it's too late. Whatever your situation, there's an abundance of knowledge, tactics, psychological insights and illuminating examples to guide you in *The 48 Laws of Power.*

But with 48 Laws to choose from and a strong possibility that any one of them might seem like a radical overhaul of your habits and thought processes, it can all seem overwhelming, impractical, or just outright impossible to put the Laws into practice. And so you put the book on the shelf and fall right back into the same old patterns. But it doesn't have to be that way.

As Robert Greene has said in many interviews, you have to choose the right Law for the right occasion. A single-minded focus on Crushing Your Enemy Totally (Law 15) isn't going to be much use if you're trying to gradually gain influence with someone, and deciding to Use Absence To Increase Respect (Law 16) will be as useful as a chocolate teapot if no one knows who you are yet. Moreover, as Robert himself notes, if you tried to practise all the Laws all the time, you'd also just be 'a horrible ugly person to be around'.

There are also contradictions between the Laws, as careful readers will have noticed. But, again, this is because we must choose and tailor the right Law for the right target and the right occasion: don't take a knife to a gun fight, and don't take a bazooka to a game of chess.

So where should we start? Which Laws should you put into practice first? And which are the most flexible and can therefore be used in almost any situation? So many questions! But help is at hand: this guide will help you take the all-important first steps on the road to Power and give you the tools you need to develop your own compass, unique to your particular life and circumstances.

And we've distilled this knowledge for you here, along with numerous, strategically-placed quotations from great minds past and present.

As well as reading and re-reading *The 48 Laws of Power* many times for our own personal development, we've spent years discussing, reflecting on, and testing each of the Laws in order to make our podcast series, Exploring The 48 Laws of Power. Andrea used the book to overcome his Autism and truly understand other people for the first time (and can now identify what makes them tick with laser-like precision). While getting his Ph.D in History, Jon used it to successfully navigate the court-like world of Oxford University and to find his way into secret archives in Russia. Along the way, we've also been able to discern the cardinal principles which underlie the Laws, which are just as (or even more) important to understand as the particular Laws you should put into practice first.

This book has two Parts. The first reveals the '3 Most Powerful Laws': the ones which you must master first and on which all the others build. Part II, 'The 4 Indispensable Power Principles', outlines the underlying concepts behind the Laws – the specific rules of thumb and social 'hacks' which explain how the Laws make sense in practice in the present day, rather than on paper, or in historical examples that can seem far removed from our modern lives.

PART

I

THE 3 MOST POWERFUL LAWS

In choosing The 3 Most Powerful Laws, our selection criteria were simple but decisive:

These Laws are the underlying principles hidden within numerous others. In other words, if you internalise these 3 you'll be well on your way to practising and understanding many more. They will be the touchstones that you can return to whenever things seem to be heading out of control.

These Laws will have the most immediate and visible impact on your life and interactions with other people. When starting any project of self-development, it's important to see immediate results to help strengthen our resolve; these Laws will deliver just that.

At the personal level, these Laws have been the ones we come back to most often in our own lives. They've rapidly become our lodestars, guiding us unerringly through the labyrinthine world of power.

So, without further ado, let's get started!

ALWAYS SAY LESS THAN NECESSARY
(LAW 3)

hen we meet other people, we want to impress – to excite their interest and curry their favour. Or we want to project an image, perhaps to intimidate and thereby impose our will. But how do we cultivate these impressions quickly and effectively? As Robert emphasises, we must hint and allude rather than be crude and direct; we must embody a sense of mystery, causing people to want to know more about us and be wary of trying to control us.

The less you say, the more people will fill in the blanks, but (and this can seem counterintuitive) they'll almost always do so to your advantage. Of course, this relies on you being selective in what you do say: if you only reveal negative or uninteresting details about yourself, people will automatically fill in more of the same. But if you merely hint at exciting projects, great achievement or expertise, they'll assume you're being modest and that the real story is 10 times more impressive. You don't need to '10X' everything; let your audience do it for you.

This is the art of the gentle smile when someone says 'Wow, that sounds really cool', or 'You must be doing really well for yourself'. The smile can mean whatever they want it to mean; talk is cheap, but the imagination is priceless.

The dangers of saying too much cannot be exaggerated. Oversharing is both burdensome to other people and it makes you appear more ordinary and unimpressive. A pithy phrase or incisive comment is infinitely more powerful in getting both attention and being treated with respect. Filling the silence with verbiage only makes you appear insecure and vacuous, so practise holding both your nerve and your tongue.

'Those who know do not speak. Those who speak do not know'

LAO TZU

The key to observing this Law is to practise not reacting to things in the moment they happen. Instead of
jumping into your familiar routines, pause. Reflect
on the situation you're in. Try to dispassionately assess
what's going on and do not immediately leap to justify,
explain or promote yourself. This pause is the true space
where both power and wisdom are cultivated. If you've
wondered why the most successful and insightful people
have a meditation practice, it's because meditation helps
to create this space between your emotional responses
arising and you actually taking action. You gain poise,
perspective, and, thereby, a great deal of power to decide
how to act rather than merely to react to other people.

Reacting emotionally – which is like reacting on auto-
pilot, blind to the specific qualities of the moment – is
a recipe for disaster. You must remain above the fray
of your own insecurities, not indulging in displays of
wounded pride or the urge to show off.

Part of this is restraining the urge most of us feel to appear significant in the eyes of others, especially if they're
people we want to befriend personally or professionally.
But you can't force this effect; on the contrary, the more
you grasp for it, the further it recedes. Like so many of
the Laws, you should think of this as a kind of seduction. You don't attract another person by talking endlessly about yourself or by over-sharing your insecurities
when you first meet. Instead, you play. You're both in a

dance – synchronising with each other's movements and remaining open to creative possibilities.

Saying less than necessary also buys you time – the most precious commodity of all. We need time to assess who we're talking to (crucial for Laws 10, 13, 19, 32, 33, among others), to understand where they're coming from, and how we can best interact with them. Becoming relaxed with a slower pace of conversation also means you won't be rushed; you subtly take control of the situation by setting the rhythm, which others will either instinctively try to match, or they will unconsciously feel they're on the back-foot.

'Listen to many, speak to few'

WILLIAM SHAKESPEARE

The less you speak, the more you can and should listen. As Robert points out in Law 29 (Plan To The End), 'So much of power is not what you do but what you do not do' – if you commit yourself too soon, you quickly forfeit control and autonomy; everyone knows the path you're on and can predict your movements (recall Law 20: Do Not Commit To Anyone). The wisest leaders listen carefully to their advisors – the experts in their respective fields – before making a decision. Louis XIV was the master of this, remaining ungraspable and unpredictable to his subordinates, as Robert tells us. But the same is true of the great boardroom leaders, fashion icons and technology trend-setters of today.

Calibrate yourself to the world and people around you, but never become predictable and mundane. You cannot Create Compelling Spectacles (Law 37) if you do not know what will seem compelling to your audience. And nor can you Discover Each Man's Thumbscrew (Law 33) if you don't listen out for his vices and insecurities. In this and countless other instances, saying less than necessary and listening to what is and is not said are habits that can't fail to help you in practising the Laws effectively.

Understanding other people's emotional needs is the secret to Power (see below), but cultivating a sharp awareness of your own emotions is just as important in this game. If you wear your hopes, fears and needs on your sleeve, you become unattractive and a burden to other people. More importantly, you also become easily manipulated by them. So, keep those hopes and fears to yourself when playing with Power: do not lean on others lest they throw you unceremoniously into the mud. Instead, through silence, listening and the art of allusion, you can rapidly get to a situation where they'll happily take you up on their shoulders.

'I begin to speak only when I'm certain what I'll say isn't better left unsaid'

CATO

*'A man's best treasure is
a thrifty tongue'*

HESIOD

2

WIN THROUGH YOUR ACTIONS, NEVER THROUGH ARGUMENT
(LAW 9)

ollowing on naturally from 'Say Less Than Necessary', Law 9 also emphasises the need to keep your mouth shut to be most effective. But this Law goes a step further: now that we've taken time to observe the situation and the person or persons we seek to influence, it's time to act.

Power is never a game which can be played alone, so we inevitably need to get other people involved in our plans. This is usually more challenging than we'd like to believe at the outset, because everyone is ultimately driven by his or her own interests, not ours. But it will take more than a few allusive words, clever arguments or even direct threats to draw their interests into line with our own. Words are simply not enough.

As the trader-turned-philosopher Nassim Nicholas Taleb puts it, 'There are two types of people. Those who try to win and those who try to win arguments. They are never the same'. Even if he exaggerates for effect (we might do both these things at different times, after all), Taleb's point is absolutely crucial.

In all walks of life, you will frequently be faced with the choice of being right or being successful. Now, this isn't about compromising your personal code or sense of honour; this is about putting aside your ego and the childish urge to publicly 'win' an argument. History is littered with stories of self-righteous individuals who refused to be practical when they could be petulant instead: perhaps their greatness is recognised after their death – like Nikola Tesla, for example – but wouldn't you rather be successful during your lifetime rather than hoping someone will eventually praise your principles when you're long gone? And even that is highly unlikely: for every Tesla, there are numberless equivalents whose stories have been entirely forgotten by History.

You must remember that trying to force someone to see things your way is an act of aggression, which everyone besides those who get off on being dominated will resent you for. So what if someone holds religious views that you find absurd, or supports a politician you find abhorrent? If you want to have influence over this person, you'll never find it by belittling their views or making them feel stupid. Think of all the times you've felt you *had* to make a point because someone else was just being too stupid for words. Did they actually ever concede the argument? And even if they did, what did that get you beyond a momentary ego-boost? Squeezing consent out of someone also more than likely leads to them resenting and undermining you in future. Never forget that bitterness is repaid far more often than kindness.

This is easy to understand. People will not help you –

and will more likely try to hinder you – if you make them feel insecure, foolish, or disrespected. It's obvious when you see it written down, but we all too easily forget it when our emotions surge forward and take control of our mouths, blind to how we're making the people around us feel. Arguing is a combat sport: someone is always going to leave the exchange bruised and bloody.

*'People can't be talked
out of illusions'*

ALAN WATTS

Metaphors aside, even if you manage to 'win' the argument – whatever it may be – people almost never change their opinions afterwards anyway. Perhaps you convince them in the moment, but they soon retreat to the comfortable familiarity of their original position, even digging in deeper to fend off any future disruption. Put simply, we like to be right... or, more accurately, to *feel* we are right. Psychological and behavioural studies have shown that we make decisions on the basis of our emotions rather than our reasoning; first we choose, and then our brains set about rationalising the decision. Knowing this, it's fruitless trying to reason people into your way of thinking – you must give them a spectacle, a demonstration, something they can feel an emotional, personal connection to – something which is beyond words and instead impacts them at a deeper, more fundamental level.

This is something every documentary- and movie-maker knows: 'show, don't tell' is the way to bring the audience along with you and, as the cliché goes, a picture paints a thousand words. And this certainly isn't a rule confined to the visual: the point is to give people a tangible story which, like one of Aesop's fables, illustrates your point in action, creating a living, breathing example for people to get their teeth into. Which is precisely what Robert himself does in his books: he uses dozens of historical examples which show us his interpretations and arguments playing out in real life. By loading his books with memorable tales and quotations from myriad different cultures,

he breathes flesh to the bones of what would otherwise
be rather abstract arguments.

*'The truth is generally
seen, rarely heard'*

BALTASAR GRACIÁN

This is backed up by research. Behavioural studies show time and again that we humans don't just love a colourful personal anecdote – we're far more likely to believe them than we are any statistical analysis. So if your friend gets mugged in a park, you'll avoid that park and experience very negative associations with it whenever it's mentioned. You won't bother checking the crime statistics to discover that the actual incidence of mugging is tiny. Nor will you give up riding your motorbike because of the stats on accidents; it takes a personal story or the death of a loved one on a bike for you to internalise that reality. In short, we're absolute pros at rejecting information until it's turned into a vivid human drama.

But let's not be black-and-white here. No doubt some people can be persuaded by logical argument alone, but, all the same, they'll then need time to process and consider this change of heart (because it will feel like a change of heart, rather than of head), so you can only sow the seeds of this change before letting them go on their way. That's a risky strategy, however, as you'll need plenty of time in hand to monitor whether they're actually coming round.

Better to tell them a story rich in colour and personality, and to tell it not as though it's your opinion, but as though it's the very evidence which convinced you. By taking them along the same path, you become fellow travellers and already begin to see things from a shared perspective.

The other side of, or key to, this Law is to always re-member that talk is cheap. From the enthusiastic inves-tor you met at some event, to your best buddy who said he'd go to the pictures with you this weekend, everybody promises things, but very few deliver. Never rely on other people coming through for you: as George R.R. Martin reminds us, 'Words are wind'; don't let that wind fill your sails, because soon enough you'll find yourself becalmed and alone. In other words, don't let yourself be swept up in the words of others: their actions are what matter, even if they do not consciously intend to deceive you.

'Tell the world what you intend to do, but first show it'

NAPOLEON HILL

The core of the dynamic at play here is that whenever we tell other people what we intend to do before we actually do it, we generate an expectation that we'll deliver. The only way you can go is down after that: they expect you will succeed, so it's not especially impressive when you do; and if you don't, well, you set yourself up for the fall. Far better to keep your mouth shut and then to deliver something out of nowhere, creating an impressive and exciting spectacle. Steve Jobs made this a speciality at Apple Keynotes, introducing 'just one more thing', like an afterthought, which usually outstripped expectations and, in that emotional sweet-spot of surprise, curiosity and novelty, thereby raised the impact of everything he'd already announced.

And that's not the only reason to keep your plans to yourself: research has shown that we give ourselves a do-pamine hit just by announcing that we'll do something, from running a marathon to writing a book. And that chemical reward for doing absolutely nothing can be addictive: people congratulating you on Facebook for saying you're going to run a marathon is in no way the same as actually running the damn thing (and you know plenty of people who announce similar 'goals' on social media which they never actually achieve).

Finally, you also need to think strategically. If you're an impresario, an entrepreneur or ideas hustler, you'll likely be working on a bunch of different projects at the same time. If you constantly talk about them, you'll rapidly (and irrevocably) become known as someone who doesn't deliver but has a big mouth. In other words, someone who's self-deluded – a label which is the kiss of death to your Power aspirations as nobody wants to support a person whose head is stuck in the clouds (or up their own ass).

So do the work and *then* deliver it. Your actions will speak for you.

RE-CREATE YOURSELF
(LAW 25)

his Law is a prerequisite for so many others, not least the 48th and final Law – to Assume Formlessness and, thereby, be able to flow effortlessly with the needs of the moment. Similarly, if you want to Pose As A Friend (Law 14), Play A Sucker (Law 21) or Act Like A King To Be Treated Like One (Law 34), you must cultivate the ability to reinvent and present different sides of yourself to the outside world. As the famous line from Shakespeare goes, 'All the world's a stage / And all the men and women merely players'… Most people think this means we each have a specific part assigned to us in the drama of life, but the truth is that we can choose and change the parts we play. We already play multiple roles, after all: a single person might be a husband, brother, musician, engineer, swimmer, traveller, friend… And each of these represents a different part of his personality. With this in mind, we can and must choose to increase and consciously develop the number of roles we can play if we are to increase our Power

'One man in his time
plays many parts'

WILLIAM SHAKESPEARE

To master this Law and its broad applications, it's particularly useful to reflect on the role of the courtier. Playing the Perfect Courtier (Law 24) is an almost forgotten art in a modern western world that emphasises individualism above all else. There's nothing wrong with forging your own unique path, but this has become misleadingly confused with the need to be the figurehead, the CEO, or the frontman of the band. Even if you do ultimately want to take that leading role, there's no shame and a great deal of power and knowledge to be gained in making yourself the power behind the throne by playing courtier – the trusted advisor, the beloved confidante – to another powerful figure.

The perfect courtier constantly shapeshifts to suit the powerful figures around them, learning to inhabit the role which the moment requires. You become the soul of re-creation, ever-fresh and exciting, making it a continual source of pleasure to have you around. Consciously practising this role will not only hone your skills in social intelligence, but will give you valuable training in the mirror effect (Law 44) as you shape yourself to either charm or disarm the people around you. The courtier remains as ungraspable as smoke... And smoke can come from sweet-smelling incense, or it could equally be the kind to choke a man to death.

'They treat me like a fox, a cunning fellow of the first rank. But the truth is that with a gentleman I am always a gentleman and a half, and when I deal with a pirate, I try to be a pirate and a half'

OTTO VON BISMARCK

It's no coincidence that one of Robert's most frequent historical examples of brilliance in the game of power is Otto von Bismarck, the 'Iron Chancellor' who outmanoeuvred all of Europe for decades and was the founding father of a united Germany in 1871. Bismarck was almost peerless in his mastery of the game of power, but he was never formally the leader of either country or army: he served his King and then his Emperor – an Emperor whom he, Bismarck, created through his political and military power-plays. No one can doubt Bismarck was the power behind the throne and behind European history in that period, but a crucial part of his success lay in his role as courtier – he could move behind the scenes, play at alliances, and remain profoundly mercurial in a way that a political figurehead who operates directly in the public eye cannot.

Like Tolkien's One Ring, Law 48's injunction to Assume Formlessness binds all the other Laws to it. And to be formless – to learn the art of flow; to be in tune with the world around you; to access the Tao – means you must always be willing and able to recreate yourself.

The very core of this principle is not to be precious about your 'identity'. The word 'authenticity' is endlessly bandied about these days, but what does it really mean when we talk about our identity? We all show different sides of ourselves to different people, consciously or otherwise. The way we behave and represent ourselves varies enormously depending on our audience, the setting, our present frame of mind, or even what we had for breakfast that morning. The 'you' you are with your closest friends is quite different to the 'you' that shows up to a business meeting, or the 'you' that takes the stage in a romantic relationship. And, of course, being an 'authentic person', a 'no-bullshit person' or whatever other kind of person, are also roles in themselves – particular characters that we choose to play. Some people find this flexibility disturbing, but there's no reason to see it that way: it's actually immensely liberating. There is simply no concrete core 'you' at all: we are free to choose how we are from moment to moment.

Buddhism, among other ancient philosophies, teaches that it's our attachment to a sense of a solid identity that is the source of all our suffering. Mastery of anything begins with mastery of the self, so we must be prepared to let go of the stories that we tell ourselves about who we are when those stories don't serve us; the freest and most powerful people in the world are those who aren't trapped in stories – neither those they tell themselves, nor the ones society foists upon them.

'In the beginner's
mind there are many
possibilities, but in the
expert's there are few'

ZEN MASTER SUZUKI ROSHI

'Stay hungry, stay foolish'

STEVE JOBS

The complementary piece of the puzzle is to remember to keep learning and never to become complacent in your position. Anything that isn't growing is decaying; a stagnant pool is the opposite of the flowing formlessness that Robert holds up as the ideal of Power. And he's certainly not the first to emphasise the need to consciously remain a student, always learning, rather than becoming self-satisfied with what we know and thus beginning to gather dust in the corner.

The beginner's mind is one of infinite possibilities, whereas the self-satisfied, self-declared 'master' has shut up shop and declared that he's learnt all he needs to. And this is not only about Power, however defined: this is about being alive. Alive to yourself, to ideas, and to the world around you.

The powerless are those imprisoned in received wisdom, who do only what they think they're 'supposed' to do, the ones who think anything new is bad and that ambitions are automatically impossible because they don't fit the cage they're in. We all grow up within such a cage – the cage of culture, social norms, the patterns we learnt from our parents – but few of us realise that the door to that cage isn't actually locked. Push it and see for yourself.

'If you ignore the rules people will, half the time, quietly rewrite them so that they don't apply to you'

TERRY PRATCHETT

'If you don't know it's impossible, it's easier to do'

NEIL GAIMAN

PART

II

THE 4 INDISPENSABLE POWER PRINCIPLES

Woven through the fabric of *The 48 Laws of Power* are four golden threads, four principles which are indispensable to your understanding of power and social intelligence. Like the points on a compass, these are the cardinal directions with which to orient yourself in the world and its power games. From the countless hours we've spent discussing and dissecting the Laws of Power on and off the air, these are the points we came back to time and again. Approach them with an open mind, because they're not quite what they seem at first glance... but first glances are what most people live by, and why most people are the victims in the continuing game of power.

THINK OF IT AS A DANCE AND A SEDUCTION

 *he 48 Laws of Power* has got a bad rap in certain circles, thanks in no small part to the publisher's decision to market it as a guide to becoming the cliché of a 'modern Machiavelli'. Boardroom execs hand out the Concise version to their fresh-faced employees in the attempt to seem cool and edgy, but they'd soon regret their showmanship if one of their recruits actually put the Laws into practice.

The reality is far more subtle. It cannot be repeated often enough that the Laws are deeply context-sensitive and should be used, not as blunt instruments, but with the careful dexterity of a surgeon's tools. Robert's work is, at the most fundamental level, about honing your social intelligence in order to better understand other people, yourself, and yourself in relation to other people. It is categorically not about using brute force or indulging in bouts of sociopathy to crash around your environment like the proverbial bull in a china shop.

What the Laws actually reveal is that when it comes to successfully interacting with other people, you must always think of it as a dance and a seduction. The dance is an act of harmonising, of bringing two different energies or melodies together to create something new. It's a duet, a back-and-forth. It's a new pattern of energy with its own dynamic that's more than the sum of its individual parts. In no way is this about drowning out other people's voices, but about finding creative ways to harmonise with them.

Remember, your voice isn't loud enough to be heard round the world all by itself. You need others to carry the message onwards, so you have to find ways to resonate with them, to have the energy ripple through them, too, extending ever-outwards like fractals. Practising how to direct that energy is what the Laws of Power are all about, but to get to that step, you need to realise that this isn't a case of playing puppet-master, but one of seduction.

And when we say 'seduction', we're talking about infinitely more than sex and romance; seduction is a dynamic which lies at the very heart of all social interactions. This is why we see the same themes and techniques emerge in books on marketing, leadership, how to make friends, pick-up and con-artistry: they have such a huge amount in common because they're all concerned with eliciting particular (positive) responses from other people.

'Everyone wants to be seduced,' says Neil Strauss in his wildly successful book, *The Game*; 'It makes us feel wanted'. But we'd go a step further than Neil and say, Everybody deserves to be seduced. They deserve to be listened to and their core need(s) understood. Forcing people to do things they don't want to is an uphill struggle and a terrible long-term strategy, so we must instead understand how they see, feel and think, and thereafter find ways to align our goals with them.

'When people talk,
listen completely. Most
people never listen'

ERNEST HEMINGWAY

Therefore, in order to be a successful seducer, you must above all be a good listener. Hear and see what the people around you are communicating and respond with these elements in mind. So many people fail to have meaningful or productive conversations because they spend their time speaking or waiting to speak – i.e. they make it all about themselves (a deeply unattractive character trait). They are neither powerful nor likeable, and that's no coincidence.

This probably isn't the image of a 'seducer' that that you first thought of when you read the word. In our sexually-repressive Western cultures, most people think that seduction is essentially A Bad Thing, but that's simply because we're usually only consciously aware of ham-fisted attempts made by inept (and therefore creepy-seeming) seducers. To completely reject seduction on this basis is like saying all music is bad when you've only heard a tone-deaf amateur try to play a guitar.

'The world wants to be
deceived'

SØREN KIERKEGAARD

*'It's still magic even if
you know how it's done'*

TERRY PRATCHETT

The true expert, a master of seduction, makes you feel extremely good about yourself and the experience you have with them. They are attentive, thoughtful and sensitive; they give you the kind of undivided attention that you long for. Like the perfect host, the perfect courtier, or the perfect date, you're made to feel wonderfully comfortable and at ease. To have someone tune themselves to your frequency rather than sitting there refusing to play in any key other than their own can be exciting as hell. There is manipulation in this, but that's not intrinsically a bad thing: this is not fraud but performance, as we've shown time and again on the podcast. To the contrary, we all crave the trick that doesn't feel like a trick.

Every sufficiently advanced form of seduction is indistinguishable from magic... and we all want some more magic in our lives.

If this still carries too many emotional connotations for you to see clearly, think back to the first time you met someone who's now a close friend. Recall the initial feeling-out process as you each cautiously showed more of yourselves to each other, rather than only your Public Face. Remember that exciting, warm feeling of gradually realising that you're on the same wavelength, but remember, too, the little tests you tried on each other to establish this: maybe it was seeing if they'd follow you into absurd or dirty humour; maybe you raised a political issue that's important to you; maybe you found something you disagreed on, but discovered that they were able to debate it without getting mad. You begin to dance; toes might get stepped on a little, but with a few adjustments, you slide into smooth synchronisation. The feeling-out process is exactly the same in seduction: it's simply practised at a more conscious and mindful level. You see The Matrix behind the interactions, rather than only doing so unconsciously and unreflectively.

The reason we find seduction or other kinds of manipulation so offensive is simply because, when it fails, it makes us suddenly aware of the process – the artifice – behind it. The magic trick is unceremoniously revealed: the hidden Aces slip from the magician's sleeve and flutter guiltily to the ground. We want the trick to work and to feel the wonder and excitement of it being faultlessly performed. But when it fails, it's a huge disappointment. Like music played badly, poor attempts at seduction are embarrassing and artificial. But when done successfully, it's an art that enriches everyone involved.

2.

TAKE CARE OF OTHER PEOPLE'S CORE
NEED FIRST

his sounds counterintuitive for a book that many read as a guide to polishing your ego. In fact, the richness of the Laws lies in their deep insights into how other people feel and respond to you. At its heart, this is a book about interpersonal dynamics, so it shouldn't really be a surprise that understanding other people is absolutely essential to practising the Laws effectively.

Ivana Chubbuck became one of the most acclaimed and sought-after acting coaches in the world by teaching the vital importance of discovering and always focusing on a character's 'Overall Objective'. This is what the charac-ter wants in life more than anything else, and which is therefore an underlying concern in every interaction they have. It might be the need for significance in the eyes of others; the need to feel certain about their life choices or to have a sense of control; it might be the need to feel excited, loved, or naughty. Whatever it is, their Overall Objective colours everything they do, and when actors are able to internalise it, to let it filter into each, smaller Scene Objective, their performances rise to an entirely new level.

But the Chubbuck Technique can be applied to real life, too, and with dramatic results. Every person you encounter is also ultimately driven by an Overall Objective that they likely have little conscious awareness of, but which nevertheless influences their every interpersonal exchange.

*'Knowing any man's
mainspring of motive
you have as it were the
key to his will'*

BALTASAR GRACIÁN

This is a theme Robert explores at length in Law 33, Discover Each Man's Thumbscrew, which is all about finding other people's weaknesses. Touching on those weaknesses can be a risky affair, but it's crucial to work out what these sensitive spots are, lest you tread on them accidentally. If someone's core emotional need is to feel respected, if you embarrass them – even accidentally – you will instantly make yourself an enemy. More importantly, despite the gruesome image, you can loosen a thumbscrew as well as tighten it, relieving the pressure, pain, and anxiety of the other person. Indeed, this may be the best tactic of all: play to their emotional need and see how they open up to you as their inner demons are tamed, at least for a while.

This is the core of the Power Principle: You must understand and work with other people's perspectives, but, above all, you must find and defuse their core emotional need.

So whatever your goal might be, before you start trying to get what you want out of any situation, stop and consider how the other person is perceiving the interaction. What do they want out of it, at the deepest level? What are they worried about? What can you do to make them feel secure, important, or appreciated? This can often be achieved quickly and simply: If they're deeply insecure, reassure them and commend their actions. If they long to feel spontaneous and adventurous, encourage them to let their hair down and act impulsively. Or if they simply long to feel important, praise and compliment them directly and watch their demeanour lighten.

'We always see our
worst selves. Our most
vulnerable selves. We
need someone else to get
close enough to tell us
we're wrong'

DAVID LEVITHAN

Sometimes it can be easier to think of this from the opposite direction: do not make other people feel insecure, embarrassed, or obligated to you. No one likes the feeling of being shown up, outdone, or that they owe someone something; this kind of crude 'power' is really no power at all: all it will breed is resentment, rejection and, ultimately, revenge.

You can certainly play on these weak points, perhaps by offering the hope of fulfilling an emotional need if they act as you want them to, or, conversely, by making it seem that their fears will come to life if they do not. Either approach requires that you observe and seek to understand the other person well in advance.

You must discard the cliché that you should treat other people as you want to be treated yourself, because this assumes that everyone experiences the world the way you do, and shares your tastes, desires, and complexes, which is patently untrue. No, you must treat people how *they* want to be treated.

The Laws help us to cultivate a deep sensitivity to other people's emotions, hopes and fears, but we must then use this information to discern their Overall Objective and act accordingly.

3.

his Principle is simple, but proves all too easy to forget if we're not careful. As we explored in Part I, we so easily get drawn into arguments instead of taking action; we brag about our big plans before we can deliver them; or we allow ourselves to be sucked into other people's cloud castles and mistake them for reality. Words certainly have power, but this is often the power to deceive.

You must never put too much weight on what other people say. They will say any number of things to advance their interests and gain favour, but most of the time it's simply hot air. If empty promises were raindrops, we'd all have drowned long ago. Of course, we often make promises with the genuine intent to follow through, but for the same reason we caution you against talking rather than putting things into action, don't fall into the trap of believing other people's words before they can show concrete evidence to back them up.

'When deeds speak,
words are nothing'

PIERRE-JOSEPH
PROUDHON

And even if they provide that evidence, don't put your trust in someone's word unless their interests clearly align with the promises they make. Unless they have a reason – positive or negative; a pull or a push – to act a particular way, then they will not do so simply because you want them to.

This isn't a dark view of human nature as being utterly selfish, however. If someone's identity is deeply entwined with a particular code of ethics, you can reasonably assume that they will follow that code even when it appears to go against their material interests. Once you uncover a person's Overall Objective, then you can also better detect how much substance lies behind their words: if their words clash with that core emotional need, even if they might feel sincere in the moment, sooner or later they'll revert back. To expect otherwise is like trying to will a compass needle to point somewhere other than North.

This Power Principle can be frustrating, even demoralising. But it does no good to sit and mope, wishing that we could believe everything we hear and not expend copious amounts of time and energy to decide whom and what to trust. All the Laws aim to describe how things really are, rather than how we – or society – might want them to be. It is that reality which you must learn to focus on, because, simply, it's the only thing which matters. It is the board we play on; we can become ever more creative and successful in how we play the game, but the board of human nature remains the same.

These days it can certainly feel that honour is in short supply, but even in the past – as Robert's many examples show – although people often spoke more about honour and the weight of one's word, things still ultimately came down to finding shared points of interest.

In short, don't be Ned Stark from *Game of Thrones*. Don't – Spoiler Alert! – get your head cut off out of sheer bloody-mindedness. Putting an idealised version of 'honour' above all practicality is actually self-indulgent ego-stroking. 'Honour' was his thumbscrew, not his strength. If you can effect no positive change for yourself, your loved ones, or your plans because of your dogmatic insistence on the letter rather the spirit and living reality of your values, then what is that honour really worth? Dogmatic inflexibility, virtuous or otherwise, is the antithesis of flow and formlessness... and really it's just moral masturbation.

'Every form of
addiction is bad, no
matter whether the
narcotic be alcohol,
morphine or idealism'

CARL JUNG

Don't get us wrong. We're not saying that you should be an amoral or an immoral person. As Detective Bunk put it in the HBO show *The Wire*, 'A man must have a code'. But to have a code, he must also still have his head. A code is only meaningful when in practice, in real life, which is never a simple affair. Be a martyr if you want, but posthumous power isn't in any sense yours to wield.

4

POWER IS RELATIONAL

ower is relational: you can only be powerful with regard to something or someone else. And you cannot be powerful without the support of other people; even an Emperor is powerless if his courtiers and warriors abandon him. So unless you want to become Lord of the Squirrels and live entirely alone in the woods, then any power must involve social interactions. Social intelligence is therefore absolutely paramount.

'Any man can fart in a closed room and say that he commands the wind'

SCOTT LYNCH

In fact, social intelligence is the unifying theme running through all Robert Greene's work. In each book he looks at different manifestations of it, but this single yet infinitely complex topic is the centre around which his work orbits. Like a sculptor, he's gradually revealed more and more of the truth trapped within the stone: his next book is explicitly dedicated to the subject and may well be his magnum opus – the final unveiling of the piece he's been crafting over the decades.

Perception is vital here: people relate to you on the basis of how they perceive you, just as your ability to accurately perceive other people's true nature – their Overall Objectives and their thumbscrews – is essential to your ability to understand and exert any influence over them.

Power is something that we invest things and people with: it is not something that exists objectively. You are important and influential only when other people think you are. It all depends on the image they hold of you, so it's imperative that you use the Laws with the conscious goal of crafting, maintaining and renewing that image. If you change people's perceptions, you change everything.

*'Power resides where men
believe it resides. No more
and no less'*

LORD VARYS
GAME OF THRONES

Many of the Laws emphasise the need to ration the time you spend in public view. Social intelligence is not just about how you talk to people, but also how regular a figure you are in their world: too much, and you become taken for granted (Laws 6, 8, 16); too little, and you are forgotten or considered suspicious (Laws 4, 18).

And this is not purely about perception; even if people consider you to be powerful, power is only meaningful when it's in motion. And this, too, requires other people. Indeed, all power is exerted through others; reputation is generated by other people sharing stories; deals are done and fortunes made by other people working on your behalf; your art and your ideas do not spread of their own volition, but only when they are shared, engaged with and passed on by others. Interpersonal relationships are the arteries through which power flows; without them, your power lies dead on the slab.

By the same token, unless you retreat to a cabin in the woods and live entirely off the land and speak to no one, you cannot opt out of the game of power by claiming that it's simply 'immoral' or 'wrong'. This in itself is a power move, as Robert points out: the people who make a show of 'opting out' publicise their stance in order 'to gain sympathy or respect', or they proclaim their 'weakness' in order to gain the power of moral authority. Even if they do so unconsciously, the effects of their actions cause ripples through the web of power relations. Never forget that you exist within this web; even if you're not conscious of it, you remain subject to its forces.

In reality, the attitude of moralising and finger-wagging masks fear and laziness. It simply feels too frightening and difficult to work to attain the goals you truly desire, so you bury them beneath a false veneer of 'realism', telling anyone who cares to listen that 'You've got to be realistic', and 'That's just how the world works', quietly suffocating your dreams beneath your excuses. This toxic passivity must be avoided at all costs, for it's another element of the Infection which Robert warns us against in Law 10.

This is why the people who speak out most vehemently against the Laws are the same kind of people who'll tell you not to stick your neck out, to toe the line and to get your head out of the clouds. Because they only see the clouds, not the open blue sky of possibilities which lies behind them. They've told themselves that the clouds are the sky and, like Asterix and his fellow Gauls, are rather afraid that at any moment it might well fall on their heads. As a social species, we're predisposed to try to fit in with the people around us – another element of relational power – so it's essential that you escape these environments or your own power will be leeched away.

All the Laws and all the other Power Principles are of no use unless you develop your social intelligence and carefully engage with the web of relational power... But make sure to do so carefully, on your own terms, and be ever-conscious of the other currents – other people's or society's power – running through the web.

REMEMBER:

The Laws are tools: they can be used positively, negatively, and every way in between

The Laws are amoral: they are neither inherently good, nor inherently bad. Robert describes how things *are*, not how they *should* be. What you do with this information is up to you

You cannot opt out of the game of power unless you want to leave society. The game continues whether you like it or not

Don't hate the player and don't hate the game. Instead, you must learn how to play and avoid being played

Perhaps the greatest power of all is intuition – the ability to understand other people and their core emotional needs. Some people are born with this capacity, but for most it's a power which we can and must cultivate

Work to develop a deep awareness of other people's emotional needs… but also your own. Do not be a slave to your emotions. Emotions and impulses arise, but we are not our emotions or impulses. We are how we choose to react to them

Always bear in mind the words of one of history's most powerful men: 'One cannot control the current of events; one can only float with them and steer' – Otto von Bismarck (which you can wear on a snazzy t-shirt, available from our website.

VOICES IN THE DARK
LEARNING HOW TO HUMAN

A Voices in the Dark Book

Be Silly. Be Kind. Be Weird.

Thank you for reading...

If you'd like, you can support our work by subscribing to our podcast, Voices in the Dark, leaving us a 5-Star review on iTunes (it gives a huge boost to the show's visibility), contributing directly, and/or telling a friend about us.

You can find and chat to us on Facebook and Twitter @VintheD, and at our home over at www.voicesinthedark.world.

CPSIA information can be obtained
at www.ICGtesting.com
Printed in the USA
BVOW11s1541060717

488535BV00007B/59/P